W9-AHT-787

BASIC/NOT BORING
SCIENCE SKILLS

SCIENCE INVESTIGATIONS

Grades 4–5

Inventive Exercises to Sharpen Skills
and Raise Achievement

Series Concept & Development
by Imogene Forte & Marjorie Frank

Exercises by Marjorie Frank

Illustrations by Kathleen Bullock

Incentive Publications, Inc.
Nashville, Tennessee

About the cover:

Bound resist, or tie dye, is the most ancient known method of fabric surface design. The brilliance of the basic tie dye design on this cover reflects the possibilities that emerge from the mastery of basic skills.

Cover art by Mary Patricia Deprez, dba Tye Dye Mary®
Cover design by Marta Drayton, Joe Shibley, and W. Paul Nance
Edited by Charlotte Bosarge

ISBN 0-86530-588-9

Copyright ©2003 by Incentive Publications, Inc., Nashville, TN. All rights reserved. No part of this publication may be reproduced, stored in a retrieval system, or transmitted in any form or by any means (electronic, mechanical, photocopying, recording, or otherwise) without written permission from Incentive Publications, Inc., with the exception below.

Pages labeled with the statement ©**2003 by Incentive Publications, Inc., Nashville, TN** are intended for reproduction. Permission is hereby granted to the purchaser of one copy of **BASIC/NOT BORING SCIENCE SKILLS: SCIENCE INVESTIGATIONS Grades 4–5** to reproduce these pages in sufficient quantities for meeting the purchaser's own classroom needs only.

1   2   3   4   5   6   7   8   9   10   07   06   05   04

PRINTED IN THE UNITED STATES OF AMERICA
www.incentivepublications.com

TABLE OF CONTENTS

Physical Science Investigations

CELEBRATE BASIC SCIENCE SKILLS

Basic does not mean boring! There is certainly nothing dull about . . .

   . . . untangling secrets of Velcro, black holes, microwaves, and ice cream.

   . . . spying on strange behaviors of plants and animals.

   . . . getting to know a few new bugs.

   . . . finding out who eats whom in the environment.

   . . . helping body parts find employment.

   . . . telling the difference between flu and food poisoning.

   . . . following speeding trains and chasing spectacular storms.

   . . . stalking shooting stars, barracudas, and diamond imposters.

These are just some of the adventures students can explore as they celebrate basic science skills. The idea of celebrating the basics is just what it sounds like—enjoying and getting good at knowing how to search for fascinating information in the world of science. Each page of this book invites learners to try a high-interest, appealing exercise that will send them tracking down information and explanations for some event or process in the natural world. At the same time, they will be sharpening their skills and knowledge in a specific area of science. This is no ordinary fill-in-the-blanks way to learn material. These investigations are fun and surprising. Students will do the useful work of deepening science knowledge while they "play detective" to answer some compelling questions.

The pages in this book can be used in many ways:

- to increase science learning and independent research for one student
- to expand the whole group's skills at locating and processing science information
- as the basis for a lesson or unit within the classroom
- by students working on their own or working under the direction of a parent or teacher

Each page may be used to introduce a new area to explore. Beyond the twenty investigations, you will find an appendix of resources helpful to the student and teacher—including a glossary of terms used in the book and a ready-to-use test for assessing the material discovered in the investigations.

The pages are written with the assumption that an adult will be available to assist the student with his or her learning. These are not science experiments, such as would be done in a lab. These are research activities. Therefore, it is essential that students have access to science resources, textbooks, encyclopedias, library books, and Internet reference sources.

As your students take on the challenges of these adventures with science facts, concepts, and processes, they will grow. As you watch them check off the science skills they have sharpened, you can celebrate with them!

The Skills Test

Use the skills test beginning on page 56 as a check-up on the understandings gained during the investigations. The test could also be used as a motivator before doing the investigations, or as a tool for adults to see how much of this material students may already know. The test can also help prepare students for further success on tests of standards, instructional goals, or other individual achievement.

SKILLS CHECKLIST FOR
SCIENCE INVESTIGATIONS, Grades 4-5

✔	SKILL	PAGE(S)
	Find and use references to locate and understand science information	10–50
	Examine and increase understanding of the system of classification of life	14–15
	Identify and explain some common plant and animal behaviors	16–19
	Identify and explain some animal characteristics	18–21
	Identify and describe some common insects and their characteristics	20–21
	Show understanding of some ecological relationships	22–23
	Identify the functions of many key human body parts	24–25
	Distinguish among common diseases and disorders of the human body	26–27
	Identify and explain some common external and internal Earth changes	28–29
	Increase understanding of the identification of minerals	30–31
	Distinguish among different kinds of storms	32–33
	Investigate mysteries and questions related to structures and living organisms in the ocean	34–35
	Identify bodies, features, and processes related to the solar system and outer space	36–37
	Increase understanding of characteristics of planets in the solar system	38–39
✔	Identify and explain examples of physics in everyday life	40–41
	Expand knowledge of key inventions and discoveries in the physical world	42–43
	Expand understanding of concepts related to force and motion	44–45
	Expand understanding of concepts related to light and color	46–47
	Expand understanding of concepts related to sound	48–49
	Explain some events caused by electricity	50

SCIENCE INVESTIGATIONS

Skills Exercises

Investigation # 1: Stalking
SEVENTEEN SCIENCE SECRETS

It will take smart snooping and sharp thinking to unravel these secrets. Search good resources for help. Try your encyclopedias, science books, other library sources, and the Internet.

When you understand the secret, explain it. Then write down your source. If you found your information on the Internet, write the web address.

1. The secret of the oyster:

How does an oyster make a pearl?

The secret revealed:_____

_____

_____

_____

_____

My source:_____

Web address:_____

3. The secret of the CD:

What device "reads" the codes and allows you to hear a CD?

The secret revealed:_____

_____

My source:_____

Web address:_____

2. The secret of the black hole:

What will happen when something gets too close to a black hole?

The secret revealed:_____

_____

_____

_____

My source:_____

Web address:_____

Investigate these secrets using your favorite Internet search engines and science sites. See page 54 for a few of the many good science Websites.

Use with page 11, 12, 13 and 54.

Name

5. The aphid's secret:

How is an aphid like a cow?

The secret revealed:_____

_____

_____

_____

_____

My source:_____

Web address:_____

4. The comet's secret:

When is the next time Halley's comet will be seen from Earth?

The secret revealed:_____

_____

My source:_____

Web address:_____

6. The diamond's secret:

What makes a diamond so hard?

The secret revealed:_____

_____

_____

_____

_____

My source:_____

Web address:_____

7. The elephant's secret:

Why does an elephant constantly move its trunk back and forth?

The secret revealed:_____

_____

_____

_____

_____

My source:_____

Web address:_____

8. The moon's secret:

Why would the moon be waxing?

The secret revealed:_____

_____

_____

My source:_____

Web address:_____

Use with pages 10, 12, 13 and 54.

Name _____

10. The microwave's secret:

When you cook food in a microwave oven, why doesn't the plate get cooked?

The secret revealed:_____

_____

_____

_____

_____

My source:_____

Web address:_____

9. The brain's secret:

How many neurons are in the human brain?

The secret revealed:_____

_____

My source:_____

Web address:_____

11. The uvula's secret:

Where can a uvula be found, and what good is it?

The secret revealed:_____

_____

_____

_____

_____

My source:_____

Web address:_____

12. The roller coaster's secret:

Why don't people fall out of a roller coaster when it goes upside down?

The secret revealed:_____

_____

_____

_____

_____

My source:_____

Web address:_____

13. The vaccination's secret:

What is the basic ingredient of a vaccination?

The secret revealed:_____

_____

My source:_____

Web address:_____

Use with pages 10, 11, 13 and 54.

Name _____

14. A shaky secret:

What moved in 1906 to cause the great San Francisco earthquake?

The secret revealed:_____

_____

_____

My source:_____

Web address:_____

15. A stormy secret:

What is a Leonid storm and when can you see one?

The secret revealed:_____

_____

_____

_____

_____

_____

My source:_____

Web address:_____

16. A sticky secret:

How does Velcro work?

The secret revealed:_____

_____

_____

_____

_____

My source:_____

Web address:_____

17. A salty secret:

Why do you need salt on the ice when you use an ice cream maker to make ice cream?

The secret revealed:_____

_____

_____

_____

_____

_____

_____

My source:_____

Web address:_____

Use with pages 10, 11, 12 and 54.

Name

Investigation # 2: Getting to Know
V.I.P.S IN THE KINGDOMS

Do they have leaves?
Do they have stems?
Do they have roots?
Where do they grow?
Can they make their own food?
How do they reproduce?
Do they have seeds?
Do they have flowers?

All living things are classified into kingdoms (large groups) and smaller groups, such as phylum, class, order, family, genus, and species. These are some of the most famous members (V.I.P.s) of the plant and animal kingdoms. (V.I.P. means Very Important Persons. Of course, most of them are not really *persons*.)

What's so special about each of these V.I.P.s? YOU find out! Search for descriptions of each member. Use the detective's questions to help your investigation, then write two or more important things about each group.

A. Algae
We're seaweed, pond weed, sea lettuce, and kelp, from the algae phylum. Here's what's unique about us:

B. Fungi
We're mushrooms, molds, and mildew—from the fungus kingdom. Here's what's unique about us:

C. Mosses & Liverworts
Here's what's special about mosses and liverworts:

D. Ferns
We're Adder's tongue and maidenhair ferns—famous ferns. Here's what's unique about ferns:

E. Seed-Bearing Plants
I'm a lovely orchid—a seed-bearing plant. Here's what's unique about seed-bearing plants:

F. Conifers
We pine trees, fir trees, cedars and junipers are proud to be conifers. Here's why:

Use with page 15.

Name

How do they look?
What special structures do they have?
Where do they live?
How do they reproduce?
Do they have a backbone?

G. Flowering Plants

We tomatoes, apples, and peaches are certainly important.
Here's what's different about flowering plants:

H. Protozoa

We're from the protist kingdom, with
relatives like a paramecium and amoeba.
Here's what is unique about us:

I. Coelenterates

Jellyfish and sea anemones like us think it's great to be
coelenterates because:

J. Mollusks

I'm an octopus. Along with snails and slugs, I enjoy being a
mollusk with these unique characteristics:

K. Annelids

I'm not just a lowly earthworm. I'm a proud annelid,
and here's what's so great about annelids:

L. Arthropods

Spiders, bees, grasshoppers, centipedes, and lobsters are all
part of our great phylum. Here's what's unique about us:

M. Echinoderms

Starfish and sea urchins are echinoderms like me, a sea
cucumber. What's special about us is:

N. Vertebrates

We fish, amphibians, reptiles, birds, and mammals in this
phylum think we're special because we have:

Use with page 14.

Name

Investigation # 3: Making Sense of
STRANGE BEHAVIORS

A curious scientist has been observing living things. Some of the behaviors she's seen are surprising indeed. Read the scientist's journal of events. Use the list on page 17 to figure out what behavior the scientist is describing.

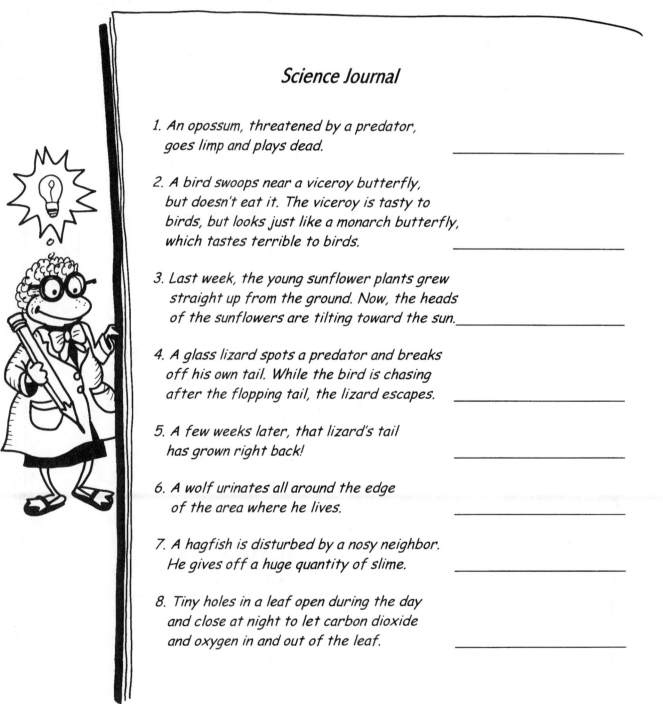

Science Journal

1. An opossum, threatened by a predator, goes limp and plays dead. _____

2. A bird swoops near a viceroy butterfly, but doesn't eat it. The viceroy is tasty to birds, but looks just like a monarch butterfly, which tastes terrible to birds. _____

3. Last week, the young sunflower plants grew straight up from the ground. Now, the heads of the sunflowers are tilting toward the sun._____

4. A glass lizard spots a predator and breaks off his own tail. While the bird is chasing after the flopping tail, the lizard escapes. _____

5. A few weeks later, that lizard's tail has grown right back! _____

6. A wolf urinates all around the edge of the area where he lives. _____

7. A hagfish is disturbed by a nosy neighbor. He gives off a huge quantity of slime. _____

8. Tiny holes in a leaf open during the day and close at night to let carbon dioxide and oxygen in and out of the leaf. _____

Use with page 17.

Name

Could it be . . .

adaptation?   dormancy?   mimicry?   reflex?
camouflage?   hibernation?   molting?   regeneration?
courtship?   instinct?   photosynthesis?   territoriality?
defense behavior?   migration?   respiration?   tropism?

9. A spider grows up knowing how to spin a web;
    nobody taught her how to do it!

    _____

10. A green plant takes carbon dioxide out of the air
     and makes sugar for its own food.

     _____

11. A woodpecker has a long, sharp bill that is just right
     for getting insects that live deep inside the tree bark.

     _____

12. A male pigeon struts around and bows
     in front of a female pigeon.

     _____

13. My dog hears the rustle of the dog food bag,
     and saliva begins dripping from his mouth.

     _____

14. The brown ptarmigan (an Arctic bird) turned
     white when the snow covered the ground.

     _____

15. A large mountain sheep successfully fights
     off every other male sheep that challenges him.

     _____

16. A male dance fly presents a dead insect
     to a female dance fly.

     _____

17. A bat eats huge amounts of food in the fall
     and gets very fat. Then he sleeps for long periods
     of time in the winter.

     _____

18. A blue whale and his family spend summers in the
     polar ocean where there is plenty of food. In the fall,
     they head back to the warm waters near the equator.

     _____

Use with page 16.

Name

Copyright ©2003 by Incentive Publications, Inc., Nashville, TN.

Basic Skills/Science Investigations 4-5

Investigation # 4: Pondering
ANIMAL ODDITIES

Odd, but intriguing!

Consider these interesting questions about animals, animal parts, or animal behavior. Research the **bold term** to find out about the animal or feature mentioned, then answer the question.

1. Is a **jellyfish** really a fish? *(If not, what is it?)* _____

2. Is a **seahorse** really a horse? *(If not, what is it?)* _____

3. Could you chop wood with a **thorax**? _____

4. Does an animal need a generator for **regeneration**? _____

5. A euglena has a **flagellum**. How is this like a flag? _____

6. Many animals have **symmetry**. Is this a kind of tree? _____

7. **Extinct** sounds like a bad smell. Is this what it means? _____

8. Might a **shark** be an animal that carts around **cartilage**? _____

9. A paramecium is surrounded with **cilia**. Are the cilia silly? _____

10. If someone has **arachnophobia**, exactly what frightens him or her? _____

11. An **exoskeleton** is not an exercise group for skeletons. What is it? _____

12. If a crocodile's **scales** are not for weighing himself, what are they for? _____

13. A squid's **tentacles** have nothing to do with camping. What are they for? _____

Use with page 19.

Name _____

This is just a bit of light reading for the weekend.

14. Is a **chrysalis** made of crystal?_____

15. Is a **slowworm** actually a worm?_____

16. How is a **scapula** like a spatula?_____

17. How is a **squid** like a fountain pen?_____

18. Why is **pride** so important to a lion?_____

19. Would you put a **sea cucumber** on a salad?_____

20. **Molt** is not the same as mold. What creature might know

    something about molt?_____

21. A beetle doesn't fly a kite with her **chitin** (pronounced KITE-un).

    So what is the chitin for?_____

22. What animal just might have the largest **proboscis**? _____

23. Where could you find a **hammer** in a human body?_____

24. **Gestation** isn't about making jokes. What is it about? _____

25. Where would you find an island (an **isle**) in the human body?_____

26. **Larva** does not come from lava. Where does it come from?_____

27. A **firefly** is not a fly. A **glowworm** is not a worm. What are they?_____

28. Would a **mud puppy** or a **water dog** enjoy a backyard doghouse? _____

29. A **marsupial** has something other animals don't have. What is it?_____

30. The largest mouse **litter** on record is twenty-four. This doesn't mean that the mouse

    left a lot of trash lying around. What does it mean?_____

Use with page 18.

Name _____

Investigation # 5: Getting Good at
GETTING BUGGED

Some people call insects "bugs," but not all bugs are the same as insects. There are many creepy, crawly creatures that are not insects, simply because they do not have all the characteristics of insects.

Professor Bea Ettle found several insects to study. She wrote a label and a description for each one. She planned to mount the insects.
But Professor Bea had two problems:
1) The insects were not dead! They were alive, and they all crawled out of their boxes.
2) Something was missing from each description.

Find the insect for each description. Write the letter of the right insect next to the description. Finish the description by finding the missing information.

> Find the common characteristics of insects. Write them here.
>
> _____   _____
>
> _____   _____

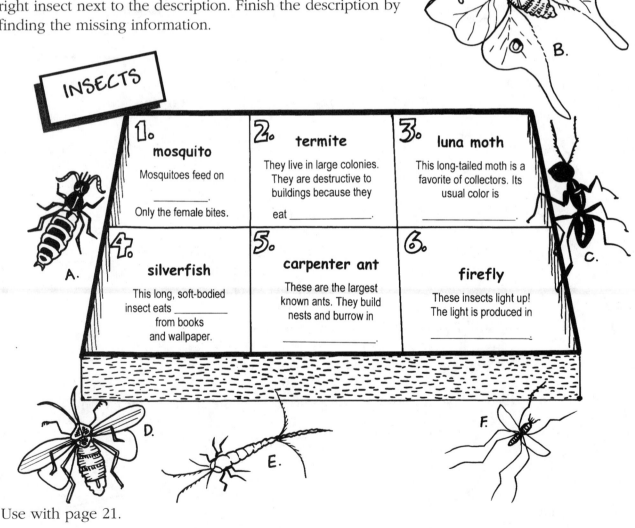

INSECTS

1. mosquito

Mosquitoes feed on _____.

Only the female bites.

2. termite

They live in large colonies. They are destructive to buildings because they eat _____.

3. luna moth

This long-tailed moth is a favorite of collectors. Its usual color is _____.

4. silverfish

This long, soft-bodied insect eats _____ from books and wallpaper.

5. carpenter ant

These are the largest known ants. They build nests and burrow in _____.

6. firefly

These insects light up! The light is produced in _____.

A. B. C. D. E. F.

Use with page 21.

Name

Copyright ©2003 by Incentive Publications, Inc., Nashville, TN.

INSECTS

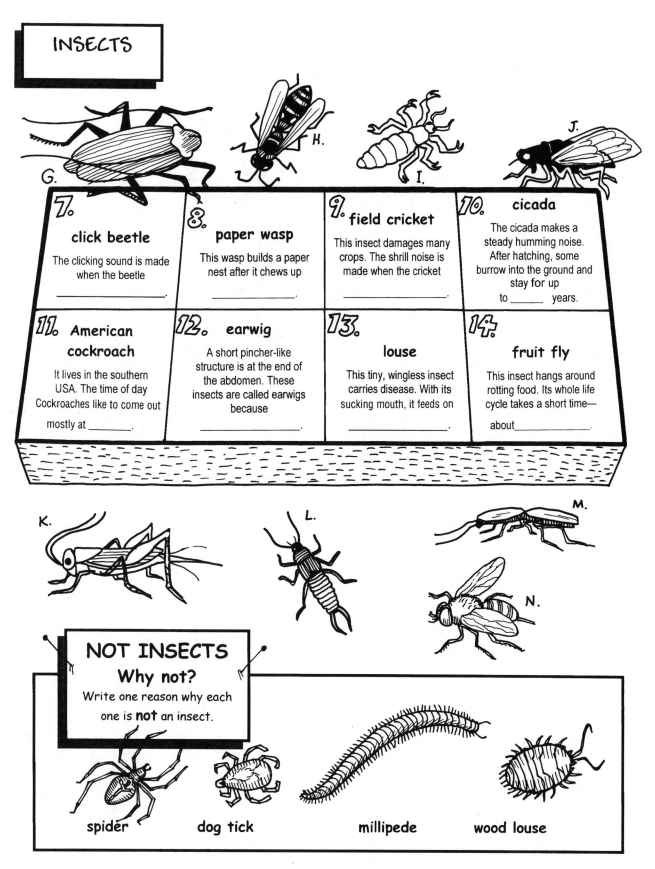

G.

H.

I.

J.

7. click beetle

The clicking sound is made when the beetle

_____.

8. paper wasp

This wasp builds a paper nest after it chews up

_____.

9. field cricket

This insect damages many crops. The shrill noise is made when the cricket

_____.

10. cicada

The cicada makes a steady humming noise. After hatching, some burrow into the ground and stay for up

to _____ years.

11. American cockroach

It lives in the southern USA. The time of day Cockroaches like to come out

mostly at _____.

12. earwig

A short pincher-like structure is at the end of the abdomen. These insects are called earwigs because

_____.

13. louse

This tiny, wingless insect carries disease. With its sucking mouth, it feeds on

_____.

14. fruit fly

This insect hangs around rotting food. Its whole life cycle takes a short time—

about_____.

K.

L.

M.

N.

NOT INSECTS
Why not?
Write one reason why each one is **not** an insect.

spider

dog tick

millipede

wood louse

Use with page 20.

Name

Investigation # 6: Figuring Out
WHO EATS WHOM?

Who eats whom?
Who eats what?

These are big questions in the environment. In the natural world, every creature is part of a food chain.

This investigation invites you to pay attention to the plants and animals in some of these food chains.

1) Make sure you understand the terms in the box below.
2) Look at the food chain pictured here.
3) Answer the questions.

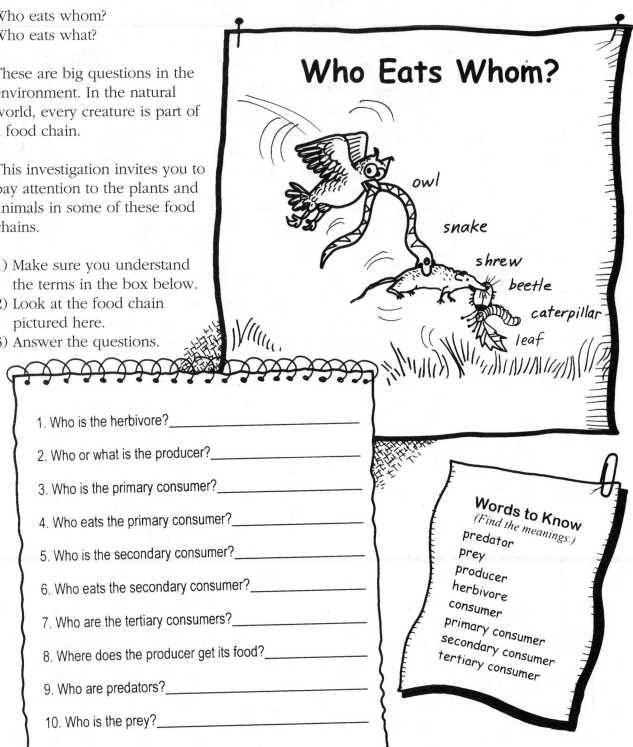

Who Eats Whom?

owl
snake
shrew
beetle
caterpillar
leaf

1. Who is the herbivore?_____

2. Who or what is the producer?_____

3. Who is the primary consumer?_____

4. Who eats the primary consumer?_____

5. Who is the secondary consumer?_____

6. Who eats the secondary consumer?_____

7. Who are the tertiary consumers?_____

8. Where does the producer get its food?_____

9. Who are predators?_____

10. Who is the prey?_____

Words to Know
(Find the meanings.)
predator
prey
producer
herbivore
consumer
primary consumer
secondary consumer
tertiary consumer

Use with page 23.

Name

 Copyright ©2003 by Incentive Publications, Inc., Nashville, TN.

11. Whom would a frog eat? _____

12. Who would eat a turtle? _____

13. Whom would a lion eat? _____

14. Who would eat a rabbit? _____

15. Whom would a pelican eat? _____

16. Who would eat an aphid? _____

17. Whom would an egret eat? _____

18. Who would eat a deer? _____

19. Whom would a lizard eat? _____

20. Who would eat a worm? _____

21. Whom would a shark eat? _____

22. Whom would a starfish eat? _____

23. Who would eat an octopus? _____

24. Whom would a rattlesnake eat? _____

25. Who would eat a zebra? _____

26. Whom would a cougar eat? _____

27. Who would eat a skunk? _____

28. Whom would a cricket eat? _____

29. Who would eat a porcupine? _____

30. Whom would an octopus eat? _____

Use with page 22.

Name

Investigation # 7: Examining
TALENTED BODY PARTS

The human body is a marvelous mix of many talented parts. Each part has a job to do, and it's important the job gets done right. Just what are those jobs? You find out!

This **Help Wanted** section of a newspaper will be advertising jobs that are available. The ad already names the body part. You need to write the job description and tell briefly what job each body part must do.

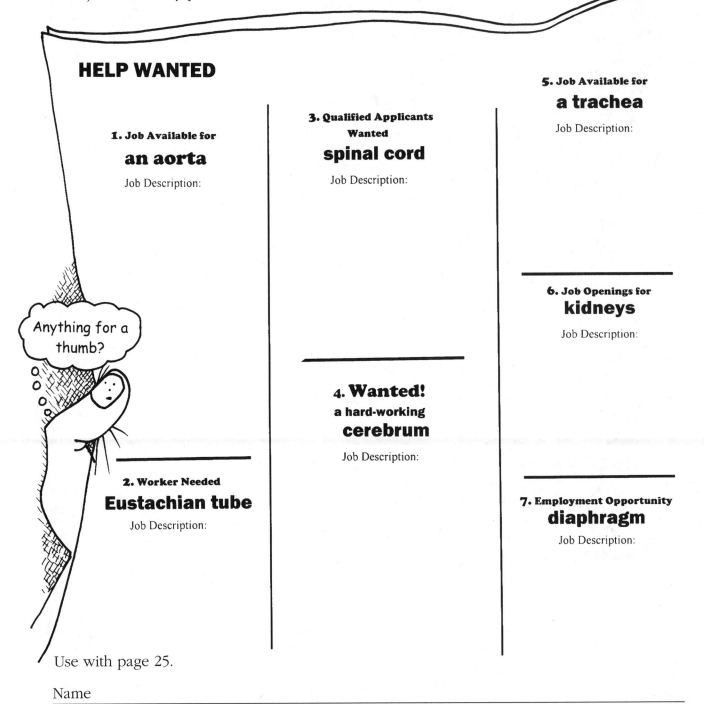

HELP WANTED

1. Job Available for
an aorta
Job Description:

2. Worker Needed
Eustachian tube
Job Description:

Anything for a thumb?

3. Qualified Applicants Wanted
spinal cord
Job Description:

4. Wanted!
a hard-working
cerebrum
Job Description:

5. Job Available for
a trachea
Job Description:

6. Job Openings for
kidneys
Job Description:

7. Employment Opportunity
diaphragm
Job Description:

Use with page 25.

Name

8. Job Available for

molars

Job Description:

9. Job Opening

tendons

Job Description:

10. Job Available for

a pair of ureters

Job Description:

11. Seeking Applicants

epidermis

Job Description:

12. Good Job Opportunity for

bronchial tubes

Job Description:

13. Needed

a thyroid gland

Job Description:

14. Positions Open for

neurons

Job Description:

15. Jobs Opening

a pupil

Job Description:

16. Jobs Available for

a stomach

Job Description:

17. Jobs Available for

a liver

Job Description:

18. Jobs Available for

white blood cells

Job Description:

19. Needed Immediately

bone marrow

Job Description:

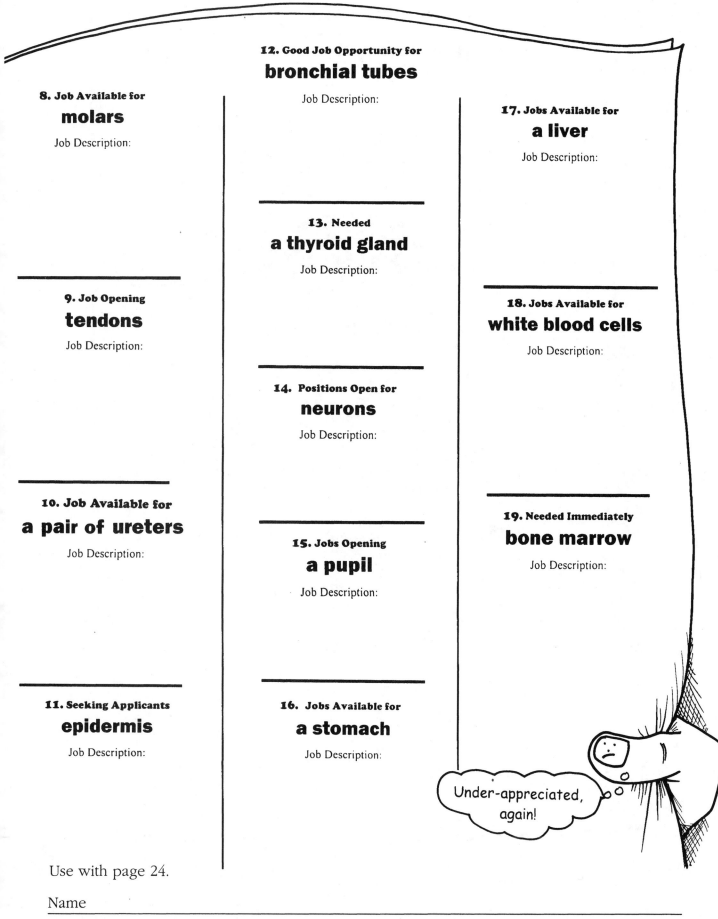

Under-appreciated, again!

Use with page 24.

Name

Investigation # 8: Identifying
CURIOUS CONDITIONS

A long line of patients is waiting to see the doctor. Each one has something disturbing to report. What conditions (or diseases or disorders) do the patients have?

Find out about each of the conditions in the doctor's *Book of Disorders*. Read the symptoms for each patient. Use the information from your research to decide what is wrong with each patient.

NEXT !

Thursday Morning Patients

Appointment Time	Patient Name	Symptoms & Description	Condition
8:30 am	J. Walker	stepped on a nail, wound infected with bacteria, nerves could be paralyzed by the toxic bacteria	
9:00 am	C.C. Spotts	mild fever and muscle aches, red spots on the face and head, blisters break open and crust forms, highly contagious	
9:30 am	Wilma Sweld	swollen, infected parotid glands, extreme tiredness	
10:00 am	Z. Z. Toad	thick, hard lumps raised from the skin surface on hands, fingers, and feet, some cracking and bleeding	
10:30 am	Elmer Socks	red, cracked skin on feet and between toes, water-filled blisters, some infected blisters on feet	
11:00 am	Beryl DeItch	thousands of dead skin cells mixed with oozing oil from hair follicles, flakes of skin on scalp and shoulders	
11:30 am	Ima Greene	extreme nausea and vomiting after eating food with a toxic germ	

Use with page 27.

Name

26 Copyright ©2003 by Incentive Publications, Inc., Nashville, TN.

Doctor's Book of Disorders

cavity
dandruff
warts
measles
tetanus
laryngitis
mumps
rabies
hepatitis
hiccups
malaria
snoring
arthritis
sprain
food poisoning
flu
bruise
chicken pox
athlete's foot
allergy
pneumonia
fracture
snoring
acne

Thursday Afternoon Patients

Appointment Time	Patient Name	Symptoms & Description	Condition
1:00 pm	G. T. Lowd	lost voice, inflammation of voice box	
1:30 pm	Angus Stiph	sore joints, frequent swelling of joints	
2:00 pm	Agatha Shiver	chills and fever, rattling sound in lungs, infection in lungs, terrible cough	
2:30 pm	Arthur Wrest	flow of air through passages at back of mouth and nose are blocked, mouth parts inside the throat vibrate with a rattling sound	
3:00 pm	Jayne Klumsie	broken blood vessels under the skin due to muscle injury, skin is discolored	
3:30 pm	Barry Blue	infection in the stomach and intestines	
4:00 pm	Erin Tayke	repeated contracting of the diaphragm, patient sucks air quickly with snapping sound, can't control the repeated event	
4:30 pm	J. J. Tripp	ankle joint twisted extremely, swelling and pain	

Use with page 26.

Name

Investigation # 9: Spying on
THE RESTLESS EARTH

Earth rarely sits still.
There are changes going on all the time.
Some are fast; some are very slow.

A patient scientist has watched Earth change for years.
She has seen many happen right before her eyes.
Other changes happen too slowly to watch,
but she observes the results of them.

Look through her telescopic lens.
What did she see happening,
or what did she notice had
happened? Write a short
description of each
Earth change.

1. moraine

2. abrasion

3. fault

4. landslide

5. delta

6. erosion

Use with page 29.

Name

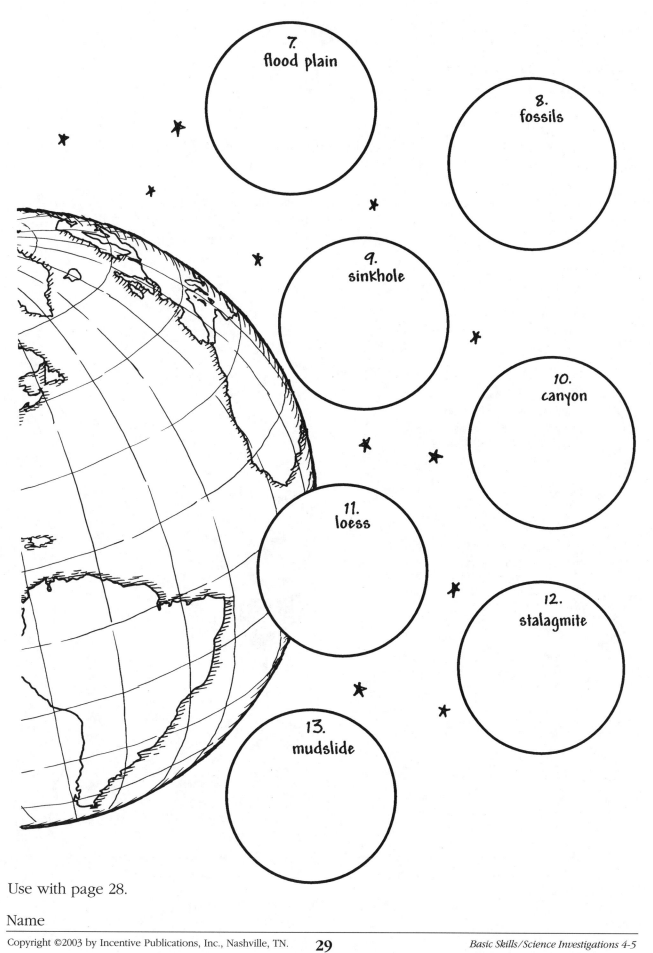

7.
flood plain

8.
fossils

9.
sinkhole

10.
canyon

11.
loess

12.
stalagmite

13.
mudslide

Use with page 28.

Name

Copyright ©2003 by Incentive Publications, Inc., Nashville, TN.

Investigation # 10: Exposing
ROCKY IMPOSTERS

Rocks are made from one or more kinds of minerals. These rocks claim to be made of certain minerals. Some of them are telling the truth about their make-up. Others are imposters (pretenders).

Find information about the properties of common minerals a copy of the hardness scale (page 55). Read the descriptions of each mineral. Could the claim be true? Find the imposters. Circle *It's for real!* or *It's an imposter!* or write the name of the mineral.

1. My name is **gypsum**. I'm a white mineral that scratches quartz.

   It's for real!   *It's an imposter!*

2. I'm **calcite**. I'm white and soft and powdery. I have a nonmetallic luster. I can be scratched by a fingernail.

   It's for real!   *It's an imposter!*

3. I'm a **diamond**. I'm the hardest known mineral on earth. I'm rare and beautiful when I'm made into jewels. I can scratch all other minerals. I am the only mineral that can scratch another diamond.

   It's for real!   *It's an imposter!*

4. My name is **sulfur**. I can scratch garnet. I'm a shiny yellow mineral.

   It's for real!   *It's an imposter!*

5. My name is **galena**. I have perfect cubic crystals. I'm a shiny, metallic gray and I make a black streak.

   It's for real!   *It's an imposter!*

6. I'm **gold!** Look at my shiny yellow nuggets. I have a hardness of 6.5 and leave a greenish black streak.

   It's for real!   *It's an imposter!*

Mystery Rock

7. My name is **halite**. My cubic crystals have a salty taste. My streak is colorless.

   It's for real!

   It's an imposter!

Wait!

8. I'm not one of those imposter minerals. I'm a beautiful yellow-red gemstone, with a hardness of 7.5. I'm a
_____.

Use with page 31.

Name _____

 Copyright ©2003 by Incentive Publications, Inc., Nashville, TN.

9. I'm **quartz**. I come in many interesting colors. I can scratch glass. I have beautiful hexagonal crystals.

It's for real!   *It's an imposter!*

10. My name is **copper**. I am coppery red. I leave a copper red streak. My hardness is 3.

It's for real!   *It's an imposter!*

11. I am the mineral **silver**. My streak is colorless. I can scratch fluorite, and I am shiny and silvery.

It's for real!   *It's an imposter!*

12. I am **topaz**. I can scratch a diamond. I'm clear and sparkly.

It's for real!   *It's an imposter!*

13. My name is **talc**. I'm soft and soapy and white. I leave a white streak. A fingernail can scratch me.

It's for real!   *It's an imposter!*

14. I am the mineral **gypsum**. A fingernail can scratch me. I'm white and I leave a white streak.

It's for real!   *It's an imposter!*

15. I'm **fluorite**. I come in many colors. My crystals are cubic. I can be scratched with a pocket knife, and I can scratch calcite.

It's for real!   *It's an imposter!*

16. My name is **feldspar**. I am the hardest mineral next to a diamond. I can scratch glass or a steel file.

It's for real!   *It's an imposter!*

17. I'm **magnetite**. I am black. I leave a black streak and I can be scratched by a penny. My crystals are cubic.

It's for real!   *It's an imposter!*

Really!

18. I am flaky and very soft. I leave a dark, greasy streak.

I'm_____.

MYSTERY ROCK II

Use with page 30.

Name

Copyright ©2003 by Incentive Publications, Inc., Nashville, TN.

Basic Skills/Science Investigations 4-5

Investigation # 11: Chasing After
SPECTACULAR STORMS

There's something fascinating about storms! These reporters are long-time storm chasers. They race around forecasting and reporting about all kinds of storms.

Find out about the storms they are forecasting here. What's so spectacular about each one? Tell what to expect from each storm. Describe the storm as clearly and completely as you can.

1. A **sand storm** is coming.
Here's what to expect:

2. Get prepared for a **blizzard**.
Here's what to expect:

3. **Sleet** is on its way.
Be ready for:

4. A **hurricane warning** is in effect.
This means that:

5. A **hurricane** has been spotted.
You can expect:

6. A **monsoon** is coming our way.
You can expect:

7. Expect **gale-force winds** today.
This means that:

Use with page 33.

Name _____

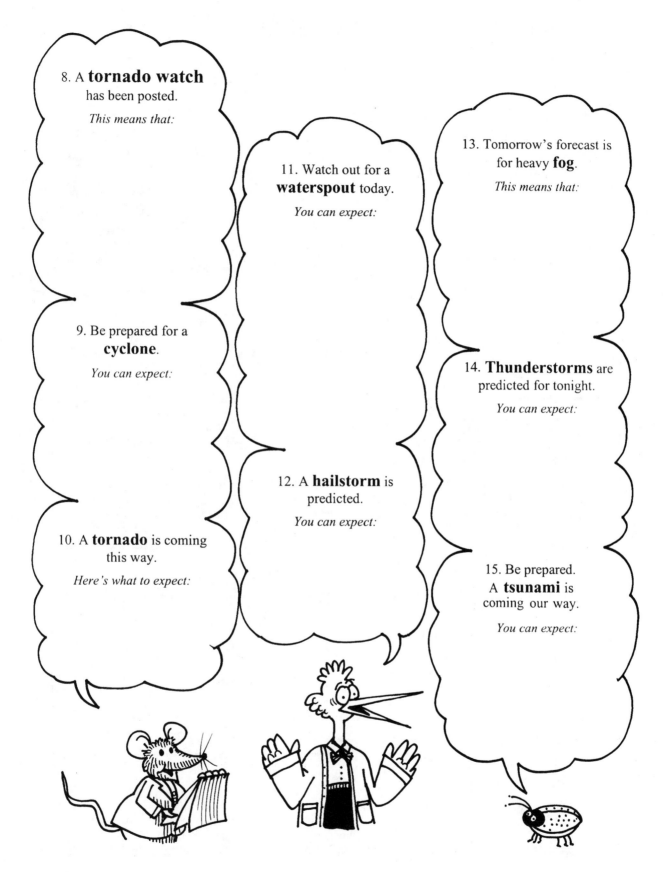

8. A **tornado watch** has been posted.

This means that:

9. Be prepared for a **cyclone**.

You can expect:

10. A **tornado** is coming this way.

Here's what to expect:

11. Watch out for a **waterspout** today.

You can expect:

12. A **hailstorm** is predicted.

You can expect:

13. Tomorrow's forecast is for heavy **fog**.

This means that:

14. **Thunderstorms** are predicted for tonight.

You can expect:

15. Be prepared. A **tsunami** is coming our way.

You can expect:

Use with page 32.

Name _____

Investigation # 12: Unraveling
MYSTERIES OF THE DEEP

The deep oceans hold many mysteries, thrills, and dangers.
There are puzzling structures, strange animals, weird plants,
and bizarre behaviors. Use the clues to help you solve each mystery.
You will probably need some help from your science textbook
or reference materials on the sea.

Mystery # 1
What is this deep, dark place?

Clue: The deepest ocean trench, in the
Pacific Ocean, is 36,000 foot canyon.

Solution:_____

Mystery # 2
What keeps the water warm?

Clue: Along the Eastern U.S. coast, the
water is surprisingly warm, even in areas of
cold climate.

Solution:_____

Mystery # 3
Where, exactly, is this strange place?

Clue: It is said that many ships, boats, and
planes have disappeared in the mysterious
Bermuda Triangle.

Solution:_____

Mystery # 4
What's the reason for this odd behavior?

Clue: Sometimes, the sea cucumber
forces out its own insides.

Solution:_____

Mystery # 5
What causes this movement?

Clue: Once or twice a day the ocean
level rises and falls. The movement is
called tides.

Solution:_____

Mystery # 6
What do they have in common?

Clue: Spits, tombolos, and barrier islands are formed from
sand which has been moved by ocean currents.

Solution:_____

Use with page 35.

Name

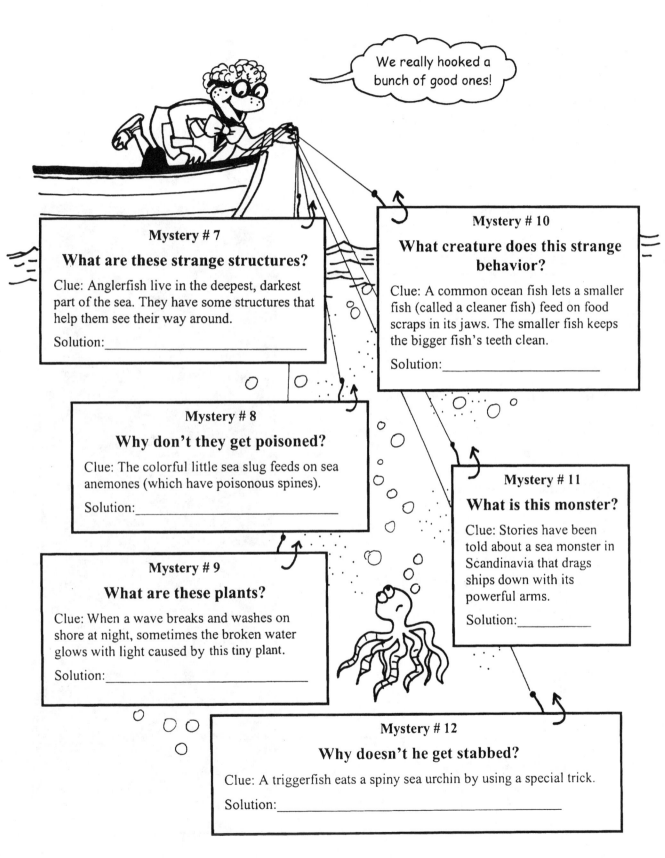

We really hooked a bunch of good ones!

Mystery # 7

What are these strange structures?

Clue: Anglerfish live in the deepest, darkest part of the sea. They have some structures that help them see their way around.

Solution:_____

Mystery # 10

What creature does this strange behavior?

Clue: A common ocean fish lets a smaller fish (called a cleaner fish) feed on food scraps in its jaws. The smaller fish keeps the bigger fish's teeth clean.

Solution:_____

Mystery # 8

Why don't they get poisoned?

Clue: The colorful little sea slug feeds on sea anemones (which have poisonous spines).

Solution:_____

Mystery # 11

What is this monster?

Clue: Stories have been told about a sea monster in Scandinavia that drags ships down with its powerful arms.

Solution:_____

Mystery # 9

What are these plants?

Clue: When a wave breaks and washes on shore at night, sometimes the broken water glows with light caused by this tiny plant.

Solution:_____

Mystery # 12

Why doesn't he get stabbed?

Clue: A triggerfish eats a spiny sea urchin by using a special trick.

Solution:_____

Use with page 34.

Name

Investigation # 13: Scrutinizing
OUTER SPACE EPISODES

Some awesome episodes take place in outer space—
throughout our solar system and beyond. Use your
investigating skills to find out about these episodes.
Describe and explain each one.

Episode A: a meteor shower

_____

_____

Episode B: a planet's rotation

_____

_____

Episode C: a planet's revolution

_____

_____

Episode D: a full moon

_____

_____

Episode E: the appearance of a comet

_____

_____

Episode F: a shooting star

_____

_____

Episode G: a meteorite

_____

_____

Use with page 37.

Name _____

Episode H: solar flares

_____

_____

Episode I: a new moon

_____

_____

Episode J: a lunar eclipse

_____

_____

Episode K: an orbiting satellite

_____

_____

Episode L: a solar eclipse

_____

_____

Episode M: a super nova

_____

_____

Episode N: a black hole

_____

_____

Episode O: a white dwarf

_____

_____

Use with page 36.

Name

Investigation # 14: Untangling
PLANETARY PECULIARITIES

Get to know the planets. Find some good references that give locations, sizes, characteristics, and movements of the planets in the solar system. Use that information to solve each *Who am I?* riddle. Write the solution to the riddle beneath the question.

1. I have more than 16 moons.
Who am I?

2. I have the largest moon.
Who am I?

3. I am the 4th planet from the sun.
Who am I?

4. I am the brightest planet as seen from Earth.
Who am I?

5. I have over 1000 rings.
Who am I?

6. I am the 7th planet from the sun.
Who am I?

7. I am the closest planet to the sun.
Who am I?

8. My size is the closest to Earth's size.
Who am I?

9. I have 10 rings and 15 moons.
Who am I?

10. It takes me 248 Earth years to make one trip around the sun.
Who am I?

Use with page 39.

Name

11. It takes me 23 hours and 56 minutes to make one rotation.
Who am I?

12. It takes me 88 Earth days to make one trip around the sun.
Who am I?

14. I am the 5th planet from the sun.
Who am I?

13. I have a moon named Charon.
Who am I?

16. I have the slowest rotation of any planet.
Who am I?

15. I have a mysterious large red spot.
Who am I?

18. When viewed from Earth, I appear red.
Who am I?

17. My orbit is between Uranus's and Pluto's.
Who am I?

20. I have no moons or natural satellites that anyone has seen.
Who am I?

19. I'm known as the "watery planet".
Who am I?

21. I am the second smallest planet.
Who am I?

22. I have ice caps at both poles.
Who am I?

Rather chatty solar system, wouldn't you say, Zerk?

Absolutely, Merk!

Use with page 38.

Name

Copyright ©2003 by Incentive Publications, Inc., Nashville, TN. *Basic Skills/Science Investigations 4-5*

Investigation # 15: Hunting for
PHYSICS IN THE NEIGHBORHOOD

Physics is the science that studies matter and energy. You can "do" physics or see examples of physics everywhere you look!

These situations are just a few examples of physics in the neighborhood. Write a sentence to explain the physics in each situation.

Look for Physics . . .

1. . . . in the kitchen.

What happens to the water molecules after the water flows into the icemaker in your freezer?

2. . . . at the deli.

There's a "whoosh" sound when you open your can of soda pop. Why?

3. . . . on the golf course.

The sound of the lawnmower is higher when the mower is coming toward you and lower when the mower has passed you. Why?

4. . . . at the grocery store.

The store checker slides each item across a glass plate when you're checking out. Why?

5. . . . at school.

The school custodian tries to pry the lid off a paint can with her fingers. She can't get it off. She finds it is easier to get the lid off when she uses a screwdriver to pry it off. Why?

6. . . . in the laundry room.

The water runs down into the laundry sink drain in a swirl. Why?

Use with page 41.

Name

7. . . . on the ski hill.

You push backward with your ski poles, but you move forward. Why?

8. . . . in the bathroom.

It is dangerous to use a hair dryer while you're taking a bath. Why?

9. . . . at the aquarium.

The big shark is a pretty heavy animal, yet he floats on top of the water and never sinks. Why?

10. . . . in a restaurant.

There is a nice, crackling fire in the restaurant's corner fireplace. You feel warmed by the fire, even though you are not touching it. Why?

11. . . . in the backyard.

Wet laundry is hung out on the clothesline in the morning. By noon it is dry. Why?

12. . . . in the garage.

You turn on the big vacuum cleaner and it sucks up all the dirt on the floor. Why does the vacuum suck up the dirt instead of blowing it around?

13. . . . at the pool.

You dive for a coin that you see on the pool bottom. When you get there, the coin is not where it appeared to be when you saw it from the surface. Why?

Use with page 40.

Name

Investigation # 16: Tracking Down
INGENIOUS INVENTIONS
& DAZZLING DISCOVERIES

Use your own ingenious, dazzling investigative skills to track down these inventions and discoveries.
Use the information in each clue to help with your research.
Write the name of the discovery or invention.

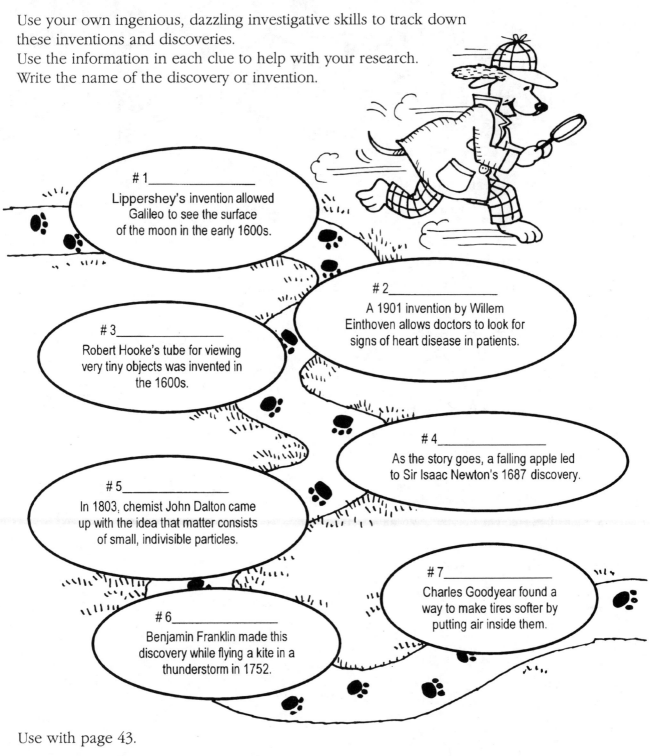

1_____
Lippershey's invention allowed Galileo to see the surface of the moon in the early 1600s.

2_____
A 1901 invention by Willem Einthoven allows doctors to look for signs of heart disease in patients.

3_____
Robert Hooke's tube for viewing very tiny objects was invented in the 1600s.

4_____
As the story goes, a falling apple led to Sir Isaac Newton's 1687 discovery.

5_____
In 1803, chemist John Dalton came up with the idea that matter consists of small, indivisible particles.

7_____
Charles Goodyear found a way to make tires softer by putting air inside them.

6_____
Benjamin Franklin made this discovery while flying a kite in a thunderstorm in 1752.

Use with page 43.

Name _____

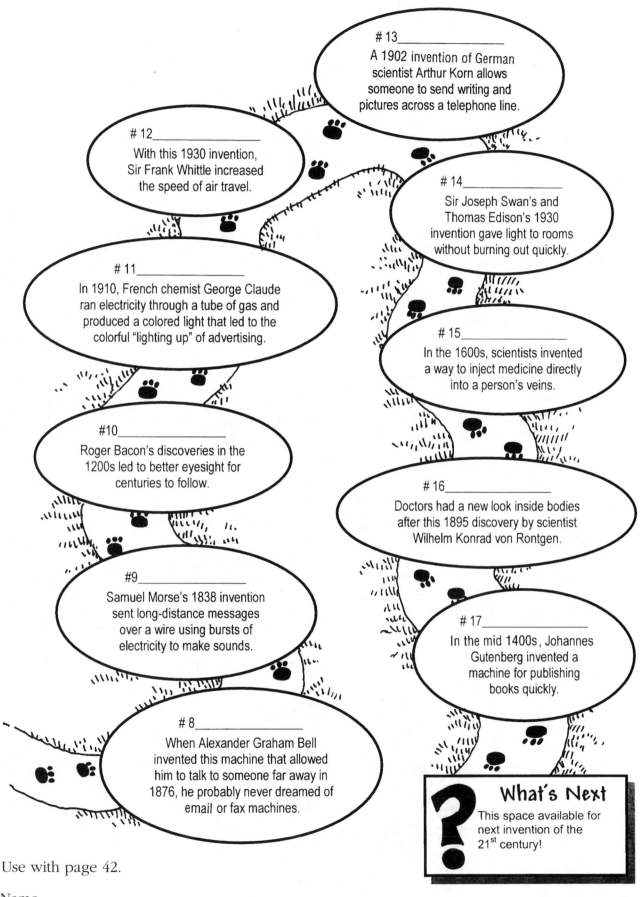

13_____
A 1902 invention of German scientist Arthur Korn allows someone to send writing and pictures across a telephone line.

12_____
With this 1930 invention, Sir Frank Whittle increased the speed of air travel.

14_____
Sir Joseph Swan's and Thomas Edison's 1930 invention gave light to rooms without burning out quickly.

11_____
In 1910, French chemist George Claude ran electricity through a tube of gas and produced a colored light that led to the colorful "lighting up" of advertising.

15_____
In the 1600s, scientists invented a way to inject medicine directly into a person's veins.

#10_____
Roger Bacon's discoveries in the 1200s led to better eyesight for centuries to follow.

16_____
Doctors had a new look inside bodies after this 1895 discovery by scientist Wilhelm Konrad von Rontgen.

#9_____
Samuel Morse's 1838 invention sent long-distance messages over a wire using bursts of electricity to make sounds.

17_____
In the mid 1400s, Johannes Gutenberg invented a machine for publishing books quickly.

8_____
When Alexander Graham Bell invented this machine that allowed him to talk to someone far away in 1876, he probably never dreamed of email or fax machines.

What's Next
This space available for next invention of the 21st century!

Use with page 42.

Name

Investigation # 17: Uncovering
NOTIONS ABOUT MOTION

A train is a good place to snoop around for notions (ideas) about motion.

Read these notions about motion. Tell if each one is correct.(Circle *correct* or *incorrect*.)
If a notion is NOT correct, write a term or fact that will fix it.

1. A train is at rest in the station. It has a kind of energy, though it is not moving. The train's conductor thinks this is **kinetic energy**.

Correct notion?　　*Incorrect notion?*

2. The train starts at a speed of 0 miles per hour. It gains speed at the rate of 3 miles per minute. In 20 minutes the train is moving at 60 miles per hour. The engineer believes that this change in speed is due to **acceleration**.

Correct notion?　　*Incorrect notion?*

3. A train runs out of fuel. Gradually the train slows to a stop without the help of brakes. A passenger on the train thinks the action of **friction** is what causes the train to stop eventually.

Correct notion?　　*Incorrect notion?*

4. The train comes to a sudden stop. But the people inside keep moving forward. This is due to **centrifugal force**.

Correct notion?　　*Incorrect notion?*

5. Two trains speed in opposite directions at the same speed, 90 mph. A passenger on one train thinks the two trains have the same **velocity**.

Correct notion?　　*Incorrect notion?*

6. A train is just about to hit a tractor that is stalled on the tracks. The tractor's owner (witnessing this safely from the side of the tracks) thinks the train will lose **momentum** and the tractor will pick up **momentum** when they collide.

Correct notion?　　*Incorrect notion?*

TRAIN DEPOT

ALL ABOARD!

Use with page 45.

Name

 Copyright ©2003 by Incentive Publications, Inc., Nashville, TN.

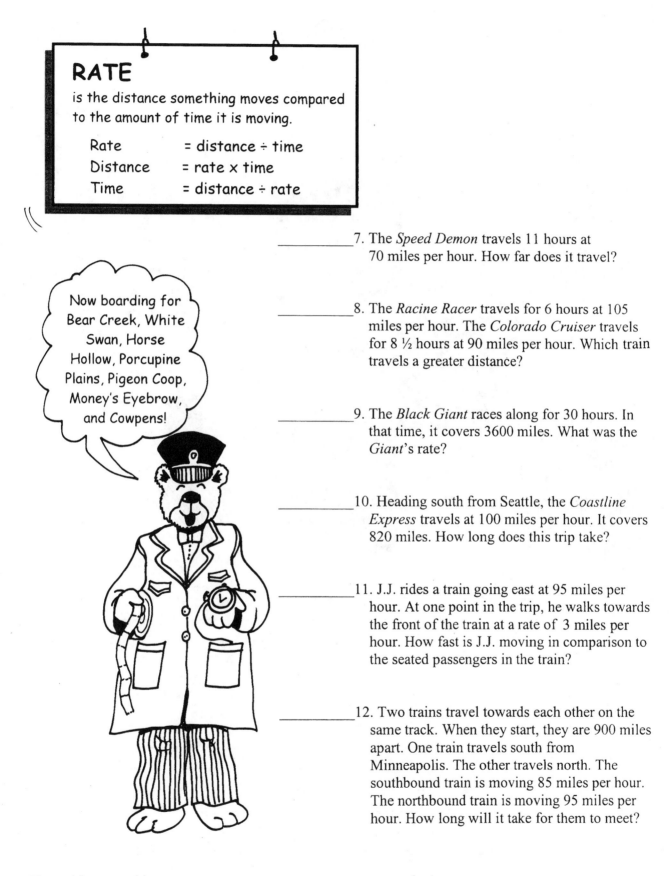

RATE

is the distance something moves compared to the amount of time it is moving.

Rate	= distance ÷ time
Distance	= rate x time
Time	= distance ÷ rate

Now boarding for Bear Creek, White Swan, Horse Hollow, Porcupine Plains, Pigeon Coop, Money's Eyebrow, and Cowpens!

_____ 7. The *Speed Demon* travels 11 hours at 70 miles per hour. How far does it travel?

_____ 8. The *Racine Racer* travels for 6 hours at 105 miles per hour. The *Colorado Cruiser* travels for 8 ½ hours at 90 miles per hour. Which train travels a greater distance?

_____ 9. The *Black Giant* races along for 30 hours. In that time, it covers 3600 miles. What was the *Giant*'s rate?

_____ 10. Heading south from Seattle, the *Coastline Express* travels at 100 miles per hour. It covers 820 miles. How long does this trip take?

_____ 11. J.J. rides a train going east at 95 miles per hour. At one point in the trip, he walks towards the front of the train at a rate of 3 miles per hour. How fast is J.J. moving in comparison to the seated passengers in the train?

_____ 12. Two trains travel towards each other on the same track. When they start, they are 900 miles apart. One train travels south from Minneapolis. The other travels north. The southbound train is moving 85 miles per hour. The northbound train is moving 95 miles per hour. How long will it take for them to meet?

Use with page 44.

Name

Investigation # 18: Explaining
TRICKS OF LIGHT

Light fools around with awesome antics and plays amazing tricks.
Learn enough about light to explain these tricks!
Write your explanations on the lines below.

1. WHAT CAUSES AN OBJECT TO LOOK BLACK?	**2.** HOW IS A RAINBOW FORMED?	**3.** WHAT CAUSES A LUNAR ECLIPSE?
4. WHAT CAUSES LIGHTNING?	**5.** WHY IS AN APPLE RED?	**6.** HOW IS LASER DIFFERENT FROM OTHER LIGHT?
7. HOW IS A MIRAGE FORMED?	**8.** WHAT CAUSES AN OBJECT TO LOOK WHITE?	**9.** WHY IS THE SKY BLUE?

1. _____
_____
2. _____
_____
3. _____
_____
4. _____
_____
5. _____
_____
6. _____
_____
7. _____
_____

Use with page 47.

Name

10. WHAT DO YOU SEE IN A KALEIDOSCOPE?	**11.** WHAT CAUSES A SUNSET?	**12.** WHY DOES A STRAW IN A GLASS APPEAR TO BE BENT?
13. WHY IS A BLACK SCREEN CALLED OPAQUE?	**14.** WHAT CAUSES A SOLAR ECLIPSE?	**15.** WHAT CAUSES AN AURORA BOREALIS?
16. WHAT MAKES A WINDOW TRANSPARENT?	**17.** WHAT CAUSES A LUMINOUS HALO AROUND THE SUN?	**18.** WHAT MAKES A WINDOW TRANSLUCENT?

8. _____

9. _____

10. _____

11. _____

12. _____

13. _____

14. _____

15. _____

16. _____

17. _____

18. _____

Use with page 46.

Name

Copyright ©2003 by Incentive Publications, Inc., Nashville, TN. *Basic Skills/Science Investigations 4-5*

Investigation # 19: Cracking
A CASE OF SOUND CONFUSION

A science detective has stumbled upon *The True Book of Sound*, by Sir Herr Plugg. This scientist has never heard of Sir Plugg, nor of the book. In fact, he can't seem to find anything about this book. Is the book reliable?

You find out how reliable this information is. Check out each statement about sound. Decide whether it is true or false. Circle the numbers of the **true** statements.

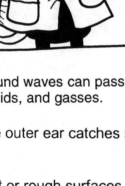

1. Sound waves can pass through solids, liquids, and gasses.

2. The outer ear catches sounds.

3. Soft or rough surfaces absorb sound.

4. An echo is reflected sound.

5. Strong vibrations produce soft sounds.

6. Sound causes the eardrum to vibrate.

7. It is dangerous for the eardrum to vibrate.

8. Weak vibrations produce soft sounds.

9. Noise is unpleasant sound.

10. Sound only travels through air.

11. Sound travels by vibrating air.

12. Sound waves cannot travel in a vacuum.

13. Sound travels fastest in liquids.

14. Sound travels at the speed of light.

15. Strong vibrations produce loud sounds.

16. Vibrations cause sounds.

Use with page 49.

Name _____

17. Some loud sounds can damage the eardrum.

18. An echo is a sound that is faster than light.

19. Pitch is highness or lowness of a sound.

20. Sound waves travel fastest through solids.

21. Sound waves cannot travel through solids.

22. Volume is the loudness or softness of sound.

23. You can speak or sing because your vocal cords vibrate.

24. Frequency is the number of times an object vibrates in a second.

25. The slower an object vibrates, the lower the pitch of its sound.

26. The faster an object vibrates, the higher the pitch of its sound.

27. To make a room quieter, cover the walls with a hard substance.

28. High-pitched sounds are made by objects that vibrate in a vacuum.

29. When a jet flies faster than the speed of sound, it causes a sonic boom.

30. People can hear sounds of things vibrating 20–20,000 times per second.

31. Dogs can hear sounds made by vibrations of more than 20,000 per second.

32. The middle ear contains bones that pass vibrations from sound to the optic nerve.

33. Sound waves travel more slowly through gases than through solids or liquids.

34. You cannot hear sounds when you're swimming underwater.

Use with page 48.

Name

Investigation # 20: Questioning
ELECTRICAL EVENTS

It seems that electricity is connected to all kinds of curious events. A curious student wonders why some things happen. He wonders why some things don't happen. Help him with his questions.

For each example, tell **why** or **why not**.

4 WHY NOT?
I flip the light switch in my closet. The light doesn't come on. Why not?

_____

_____

1 WHY?
Every time I comb my hair with a plastic comb, my hair crackles and stands up on end. Why?

_____

_____

5 WHY?
I just ran across the carpet in my socks (no shoes). When I reached out to touch the metal doorknob, I got quite a shock. Why?

_____

_____

2 WHY NOT?
A family was riding in a car. Lightning struck the car but no one inside was harmed. Why not?

_____

_____

6 WHY NOT?
During a fierce lightning storm, lightning struck a tree in our yard. It did not strike the ground next to the tree. Why not?

_____

_____

3 WHY?
I see the soccer ball coming toward me, fast. My foot jerks out to kick the ball. Why?

_____

_____

7 WHY?
We just got a new electric water heater. It is all wrapped in a puffy covering. Why?

_____

_____

Name _____

50 Copyright ©2003 by Incentive Publications, Inc., Nashville, TN.

APPENDIX

CONTENTS

TERMS FOR
SCIENCE INVESTIGATIONS

acceleration — the rate at which velocity changes

adaptation — a trait in an organism that allows it to change in order to survive

caldera — the large sunken valley resulting when a volcano erupts

camouflage — body coloring or change in body coloring to protect an animal

cavern — a large underground chamber or cave

centripetal force — the force on an object acting toward the center of a circular path

comet — a mass of frozen gases, dust particles, and rock particles that orbits the sun

consumer — an organism that eats other organisms

courtship — the behaviors leading up to mating

decomposer — an organism that causes the decay of dead organisms

defense behavior — behaviors used by an organism to help it survive or escape from danger

delta — a fan-shaped deposit that develops when moving water loses its velocity

dormancy — a period of time when the plant embryo is resting

eclipse — the passing of one object into the shadow of another object

erosion — the processes by which Earth materials are worn away or carried away and deposited

exoskeleton — the hard outer covering that protects the internal organs of an arthropod

extinction — the total disappearance of a species

fault — a fracture in a rock along which movement has taken place

flood plain — an area of fertile soil composed of fine sediment deposited during floods

fossil — the remains or traces of once-living organism preserved in rocks

friction — a force that opposes motion between two surfaces that touch each other

gestation — the period of time between fertilization and birth

herbivore — an animal that eats only plants

hibernation — an adaptation for winter survival in which an animal's body functions slow down

inertia — the property of a body that resists any change in velocity

instinct — an inborn behavior that involves complicated responses to a stimulus

kinetic energy — energy of motion

kingdom — the major classification category of living organisms

landslide — the rapid downhill movement of large amounts of Earth materials

loess — a windblown deposit of fine dust particles from deserts, dry riverbeds, and old glacial lakebeds

meteor — meteoroids that burn up in Earth's atmosphere

meteorite — small fragments of meteor material that strike Earth's surface

migration — the movement of animals or groups of animals

mimicry — a behavior where an animal copies another organism in order to protect itself

mineral — a naturally occurring, inorganic, crystalline solid with a definite chemical make-up

momentum — the product of an object's mass and its velocity

moraine — ridges of debris deposited by melting glaciers

mudslide (or mudflow) — rapid, downhill movement of materials occurring after heavy rains

neuron — nerve cell; body cell that carries electrical impulses sent from the brain

opaque — material that absorbs light

photosynthesis — the process by which plants use light energy to make food

phylum — the largest classification category in a kingdom

physics — the science of matter and energy

pitch — a quality of sound (highness or lowness) determined by wave frequency

potential energy — energy due to position or condition

predator — animal that captures another animal for food

prey — the animals eaten by a predator

primary consumer — an animal that eats a producer

producer — organism that contains chlorophyll to make food by photosynthesis

rate — the distance moved in relation to the time taken for the movement

respiration — the process by which cells use energy from food for their life activity

reflex — a quick, automatic response to a stimulus that does not involve the brain

regeneration — regrowing lost body parts

revolution — the movement of a body (or object) around another body (or object)

rotation — the turning or spinning of an object on an axis

secondary consumer — the animal who eats the animal that eats the producer

sinkhole — a funnel-shaped depression in Earth's surface that results from limestone being dissolved

species — the smallest category in the kingdom in which only one kind of organism is classified

spit — a long ridge of sand deposited by a longshore current when its velocity slows

stalactite — an upward growth of calcium carbonate from the floor of a cave

stalagmite — an elongated structure of calcium carbonate that hangs from cave ceilings

static electricity — electricity produced by charged bodies; charge built up in one place

territoriality — animal behavior that acts to protects its area

tertiary consumer — a consumer who eats an animal that is a secondary consumer

tides — shallow water waves caused by the gravitational attraction among Earth, moon, and sun

translucent — light passes through it but you cannot see through it

transparent — light passes through it and you can see through it

tropism — the response of a plant to a stimulus

tombolo — a sand or gravel deposit that connects an island to the mainland

vaccine — a solution of dead or weakened germs that causes immunity against a disease

velocity — the speed and direction of a moving object

vibrations — rapid back and forth movements

waning — getting smaller

waxing — getting larger

SCIENCE WEBSITES TO EXPLORE

These Internet websites are some of the sites you might try to help you unlock the science secrets on pages 10–13. Try other favorite science sites and non-Internet sources, too.

American Lung Association—www.lungusa.org/

Aqua Network-Aquatic World—www.aquanet.com

Ask Dr. Universe—www.wsu.edu/DrUniverse

The Aurora Page—www.geo.mtu.edu/weather/aurora

Beakman & Jax—www.beakman.com

Biology Learning Center–Marine Biology—www.marinebiology.org/science.htm

Cool Science for Curious Kids—www.hhmi.org/coolscience

Discovery Channel Online—www.discovery.com

For Kids Only—Earth Science Enterprise—kids.earth.nasa.gov

The Exploratorium—www.exploratorium.edu

Fear of Physics—www.fearofphysics.com

NASA Kids—kids.msfc.nasa.gov/

Neuroscience for Kids—faculty.washington.edu/chudler/neurok.html

National Air and Space Museum—www.nasm.si.edu

National Institute of Environmental Health Services—www.niehs.nih.gov/kids/home.htm

National Science & Technology Week for Kids—www.nsf.gov/od/lpa/nstw/kids/start.htm

Oceanlink—www.oceanlink.island.net

Rader's Biology 4 Kids—www.biology4kids.com/

Rader's Chemistry 4 Kids—www.chem4kids.com

Rader's Geography 4 Kids—www.geography4kids.com

Reeko's Mad Science Lab—www.spartechsoftware.com/reeko

Rob's Granite Page—uts.cc.utexas.edu/~rmr/index.html

Science 4 Kids, Agricultural Research Service of the USDA—www.ars.usda.gov/is/kids/

Science Made Simple—www.sciencemadesimple.com

Sea & Sky—www.seasky.org

Smithsonian Museums—www.si.edu

Whale Songs—www.whalesongs.org

Yuckiest Site on the Internet—yucky.kids.discovery.com

NOTE: The Internet changes daily. Websites recommended here have been chosen carefully; however, a teacher or parent should review any site before directing students to the site.

PHYSICAL PROPERTIES
OF SOME COMMON MINERALS

Metallic Luster

Mineral	Color	Streak	Hardness	Crystals
GRAPHITE	black to gray	black to gray	1-2	hexagonal
SILVER	silvery, white	light gray to silver	2.5	cubic
GALENA	gray	gray to black	2.5	cubic
GOLD	pale golden-yellow	yellow	2.5-3	cubic
COPPER	copper red	copper red	3	cubic
CHROMITE	black or brown	brown to black	5.5	cubic
MAGNETITE	black	black	6	cubic
PYRITE	light brassy yellow	greenish black	6.5	cubic

Nonmetallic Luster

Mineral	Color	Streak	Hardness	Crystals
TALC	white, greenish	white	1	monoclinic
GYPSUM	colorless, gray, white	white	2	monoclinic
SULFUR	yellow	yellow to white	2	orthorhombic
HALITE	colorless, red, white, blue	colorless	2.5	cubic
CALCITE	colorless, white	colorless, white	3	hexagonal
DOLOMITE	colorless, white, pink, green, gray	white	3.5-4	hexagonal
FLOURITE	colorless, white, blue, green, red, yellow, purple	colorless	4	cubic
HORNBLENDE	green to black	gray to white	5-6	monoclinic
FELDSPAR	gray, green, white	colorless	6	monoclinic
QUARTZ	colorless, colors	colorless	7	hexagonal
GARNET	yellow-red, green, black	colorless	7.5	cubic
TOPAZ	white, pink, yellow, blue, colorless	colorless	8	orthorhombic
CORUNDUM	colorless, blue, brown	colorless	9	hexagonal
DIAMOND	colorless, dark, many colors	colorless	10	octagonal or hexagonal

Moh's Hardness Scale

Mineral	Hardness	Hardness Test
TALC	1	softest, can be scratched by a fingernail
GYPSUM	2	soft, can be scratched by a fingernail but cannot be scratched by talc
CALCITE	3	can be scratched by a penny
FLOURITE	4	can be scratched by a steel knife or a nail file
APATITE	5	can be scratched by a steel knife or nail file, but not easily
FELDSPAR	6	knife cannot scratch it; it can scratch glass
QUARTZ	7	scratches glass and steel
TOPAZ	8	can scratch quartz
CORUNDUM	9	can scratch topaz
DIAMOND	10	can scratch all others

Copyright ©2003 by Incentive Publications, Inc., Nashville, TN. *Basic Skills/Science Investigations 4-5*

SCIENCE INVESTIGATIONS
SKILLS TEST

Each answer is worth 1 point. Total possible points = 60.

1–9. Write T for each true statement and F for each false statement.

____ 1. A black hole has a very strong gravitational pull.

____ 2. Comets revolve around Earth.

____ 3. The moon is waxing when is moves from a new moon to a full moon.

____ 4. Gravity is the force that keeps you from falling out of an upside-down roller coaster.

____ 5. Fungi are seed-bearing plants.

____ 6. Starfish and sea cucumbers are fish.

____ 7. Anemones and sea urchins are mollusks.

____ 8. A firefly is really a beetle.

____ 9. People and alligators are vertebrates.

10–19. Write one answer (or term) for each statement. You may use one from the lists shown, but your answer does not have to come from these lists. Other answers may be correct.

_____ 10. This plant belongs to the algae phylum.

_____ 11. This animal is an arthropod.

_____ 12. This is a segment of an insect's body.

_____ 13. A predator breaks off a starfish's arm. The arm grows back.

_____ 14. A snake grows too big for its skin and sheds it.

_____ 15. A chameleon turns green while resting on a green leaf.

_____ 16. A spider knows how to spin a web without being taught.

_____ 17. A squid squirts ink at a predator.

_____ 18. Dust blows up your nose and you sneeze.

_____ 19. A seed rests for a long time after fertilization before sprouting a plant.

Plants & Animals

sequoia	moss	sea kelp
octopus	fern	sponge
onion	shrimp	thorax
spider	fruit fly	antennae

Behaviors

adaptation
camouflage
defense
dormancy
hibernation
instinct
migration
mimicry
molting
photosynthesis
reflex
regeneration
respiration
transpiration

Name

FOOD CHAIN

FOX → EAGLE → SHREW → BEETLE → CATERPILLAR → LEAF

20–25. *Use the names of the organisms in the picture to answer the questions.*

20. Who is the producer in the chain? _____

21. Who is the primary consumer? _____

22. Who is the secondary consumer? _____

23. Name one predator in the chain. _____

24. Name one prey in the chain. _____

25. Name one tertiary consumer in the chain. _____

26–32. *Circle one or more correct answers for each question.*

26. Which of the following are *not* insects?
    a. firefly   b. paper wasp   c. tarantula   d. termite   e. centipede   f. tick   g. earwig

27. Which of the following are primary consumers?
    a. deer   b. lion   c. caterpillar   d. fox   e. cow   f. rabbit   g. owl

28. Which body part(s) filters impurities out of the blood?
    a. aorta   b. spinal cord   c. liver   d. stomach   e. bone marrow   f. tendons

29. Which body part(s) have jobs that are not related to breathing?
    a. ureters   b. bronchial tubes   c. diaphragm   d. trachea   e. molars   f. kidneys

30. Which ailment(s) would affect the skin or show on the skin?
    a. warts   b. bruise   c. laryngitis   d. food poisoning   e. pneumonia   f. athlete's foot

31. Which minerals are harder than graphite?
    a. talc   b. topaz   c. diamond   d. calcite   e. corundum   f. quartz

32. Which mineral(s) would scratch corundum?
    a. silver   b. copper   c. corundum   d. diamond   e. talc   f. gold

Name _____

33–39. *Write the letter of the Earth change that fits the description.*

____ 33. the rapid downhill movement of large amounts of Earth materials

____ 34. scouring action of wind-carried particles resulting in erosion of rock

____ 35. a funnel-shaped depression in Earth's surface caused by dissolving limestone

____ 36. an upward growth of calcium carbonate from the floor of a cave

____ 37. an area of fertile soil composed of fine sediment deposited during floods

____ 38. a fan-shaped deposit that develops when moving water loses its velocity

____ 39. the processes by which Earth materials are worn away or carried away and deposited

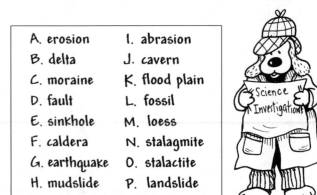

A. erosion	I. abrasion
B. delta	J. cavern
C. moraine	K. flood plain
D. fault	L. fossil
E. sinkhole	M. loess
F. caldera	N. stalagmite
G. earthquake	O. stalactite
H. mudslide	P. landslide

40–44. *Write an answer to finish each sentence.*

_____ 40. is the ocean current that keeps water warm along the eastern U.S. coast

_____ 41. is the planet closest to the sun.

_____ 42. moves into Earth's shadow during a lunar eclipse.

_____ 43. is the 8th planet from the sun.

_____ 44. is the planet with over 1000 rings.

51. Explain the difference between these: a tornado, a hurricane, and a typhoon.

45–51. *Circle the true statements.*

45. A monsoon is a tornado over water.

46. Gale force winds are between 15 and 50 MPH.

47. A tsunami is not a storm.

48. Tides are caused by the gravitational pull of the sun, moon, and Earth.

49. A shooting star is a tiny comet.

50. All planets rotate at the same speed.

51. *Answer the question in the box.*

_____

_____

_____

Name

Basic Skills/Science Investigations 4-5   **58**   Copyright ©2003 by Incentive Publications, Inc., Nashville, TN.

52–55. *Give a short explanation for each question.*

52. Why is it dangerous to use a hairdryer while taking a bath?

_____

53. Why do people get in a car for safety during a lightning storm?

_____

54. A skateboarder pushes backwards with his left foot. Why does the skateboard go forward?

_____

55. Why does an object appear to be black?

_____

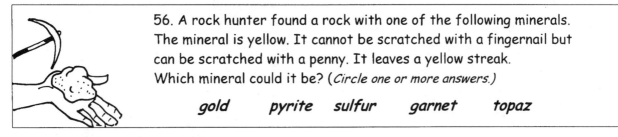

56. A rock hunter found a rock with one of the following minerals. The mineral is yellow. It cannot be scratched with a fingernail but can be scratched with a penny. It leaves a yellow streak. Which mineral could it be? (*Circle one or more answers.*)

gold   pyrite   sulfur   garnet   topaz

57–59. Circle one or more correct answers.

57. Which statements are true?
   a. Benjamin Franklin discovered gravity in a thunderstorm.
   b. Jet engines were not invented until the 1960s.
   c. An electrocardiogram is an invention that helped find heart disease.
   d. Guttenberg's invention helped speed up the printing of books.

58. Which statements are true about sound?
   a. Sound waves cannot travel in a vacuum.
   b. Frequency is the number of times an object vibrates in a second.
   c. Strong vibrations produce soft sounds.
   d. People can hear sounds at higher frequencies than dogs.

59. Which statements are true?
   a. Kinetic energy is the energy of motion.
   b. Acceleration is increase in velocity over time.
   c. Friction slows down moving vehicles.
   d. Two trains going opposite directions at the same speed have the same velocity.

60. Solve the problem by finding the rate.

60. A speedy train, *The Black Dragon,* left the station at 9:00 p.m. Pacific Time. It arrived at its destination 665 miles away at 4:00 a.m. the next morning (same time zone).

At what rate was *The Black Dragon* traveling?

Answer_____

Name _____

SKILLS TEST ANSWER KEY

1. T
2. F
3. T
4. F
5. F
6. F
7. F
8. T
9. T

10–19: Answers may vary. Students may use other answers for these that are not shown on the chart.

10. sea kelp
11. shrimp, spider, or fruit fly
12. thorax
13. regeneration
14. molting
15. camouflage
16. instinct
17. defense
18. reflex
19. dormancy
20. leaf or plant
21. caterpillar
22. beetle
23. fox, eagle, shrew, or beetle
24. eagle, shrew, beetle, or caterpillar
25. shrew, eagle, or fox
26. c, e, f
27. a, d, e, f
28. c
29. a, e, f
30. a, b, f
31. b, c, d, e, f
32. c, d
33. P or H

34. I
35. E
36. N
37. K
38. B
39. A
40. Gulf Stream
41. Mercury
42. the moon
43. Neptune
44. Saturn

45–50: 47 & 48 are true (should be circled).

51. A tornado is a violent, whirling storm that moves in a narrow path over land; a hurricane is a warm, moist storm that develops over tropical waters; a waterspout is a tornado over water.

52–55: Answers will vary. Give credit for any answer which gives a reasonable explanation showing understanding of the science involved.

52. Water is a good conductor of electricity (and will bring the electricity to your body!)
53. The car's rubber tires do not conduct electricity well.
54. For every action there is an opposite reaction.
55. Black does not reflect any of the colors in light.
56. gold or sulfur
57. c, d
58. a, b
59. a, b, c
60. 95 mph

ANSWERS TO INVESTIGATIONS

Answers to the secrets may vary some, but should contain the substance of each of these below:

1. oyster—An oyster has a pearly layer that lines its shell, formed by cells inside the shell. When a foreign substance gets into the oyster, these cells produce a substance called nacre and cover the foreign object with layers of nacre until it is enclosed in a "shell" called a pearl.
2. black hole—Anything close to a black hole will get drawn in and crushed by the tremendous force of gravity in the hole.
3. CD—a laser beam
4. comet—the year 2062
5. aphid—Many aphids live with ants. They eat plant juices and get fat. The ants "milk" the aphids to get the juice from them for food.
6. diamond—It's made of carbon whose atoms are in a structure where they are very tightly held together.
7. elephant—The elephant is using its sense of smell, constantly sniffing the air for danger.
8. moon—when it is traveling around Earth from new moon to full moon phase. This path allows more and more of the sun's rays to be reflected from it, making the moon more visible from Earth. (Waxing means the moon is getting bigger.)
9. brain—100 billion neurons
10. microwave—The plate is made from a material that doesn't absorb energy from the oven's magnetic field the way the food does.
11. uvula—It's in the back of the throat. It helps to form particular sounds by its vibrations.
12. roller coaster—The force of inertia that keeps the body moving forward along the track is greater than the force of gravity pulling down on their bodies.
13. vaccination—Vaccines are made from dead or weakened germs.
14. earthquake—the San Andreas fault
15. Leonid storm—is a trail of pebbles and dust from a passing comet (Tempel-Tuttle) that causes a spectacular meteor shower. It can be seen each year in mid-November.
16. Velcro—One side is a collection of tiny hooks. The other is tiny loops. The hooks catch onto the loops and hold fast.
17. salt—The salt lowers the temperature of the ice as it melts, making the ice cold enough to help the ice cream freeze.

Answers will vary. Students will choose different characteristics to emphasize. The answers below are guidelines. Check for accurate, general understanding.

A. very simple plants, no stems, no roots, no leaves, grow in wet places, make their own food, nonvascular
B. many-celled organisms, not green, do not make food, absorb food from other organisms (mostly dead or dying), produce spores
C. thin leaves, many cells, nonvascular, make their own food, reproduce by spores, live in wet areas
D. green, make food, have stems, leaves and roots, grow in moist places, reproduce with spores, have feathery leaves called fronds
E. vascular, have roots, stems and leaves, reproduce with seeds, make food, some produce flowers
F. seed-bearing plants, seeds produced in cones, leaves are in the shape of needles, green, make food
G. seed-bearing plants, seeds produced inside a fruit which develops from a flower, make own food
H. kingdom of one-celled or very simple organisms, usually live in wet places, most take in food, some move by means of cilia, flagella, or pseudopods
I. animal with body around a central cavity, most have tentacles, live in water
J. soft-bodied animals with shells, have a single strong foot for movement, live in water or on land
K. worms with segmented bodies, have bristles, live in water or on land
L. have an exoskeleton and segmented bodies, many have antennae, bodies usually divided into three segments: head, thorax, abdomen, four main classes are crustaceans, insects, arachnids, and myriapods
M. spiny animals, most live in salty water, have radial symmetry
N. animals with backbones, have specialized body systems, main classes are birds, reptiles, amphibians, mammals, and fish

Answers may vary somewhat. More than one behavior may fit a description.

1. defense behavior
2. mimicry
3. tropism
4. defense behavior
5. regeneration
6. territoriality
7. defense behavior
8. respiration
9. instinct
10. photosynthesis
11. adaptation
12. courtship
13. reflex
14. camouflage
15. dominance
16. courtship
17. hibernation
18. migration

1. No, it is a coelenterate.
2. No, it is a small fish.
3. No, a thorax is a part of an insect or other arthropod.
4. No
5. A flagellum waves back and forth.
6. No
7. It means that something no longer exists.
8. Yes
9. No
10. spiders
11. a hard outer covering that protects the inside parts of an arthropod
12. They prevent loss of water from the body.
13. They are used to obtain food.
14. No
15. No it's a lizard without feet.
16. It is flat and strong. (Answers may vary.)
17. It can produce (or squirt) ink.
18. A pride is the group or family in

which a lion lives.
19. no
20. a snake or other animal that might shed its skin
21. It forms the exoskeleton of arthropods.
22. an elephant (or any other animal with a long nose)
23. in the ear
24. It is the period between conception and birth.
25. in the pancreas (Isles of Langerhans)
26. Larva hatches from the egg of an insect.
27. beetles
28. no
29. a pouch for carrying newborns
30. A litter is a group of newborn babies.

pages 20–21

Insect Characteristics: segmented body (3 segments), 3 pairs of legs, most have antennae and wings, arthropod
1. blood from a human or other animal
2. wood
3. green
4. starch
5. wood (or dead wood)
6. lower abdomen
7. flips over in the air
8. wood
9. rubs its forewings together
10. 20 years (answers may vary)
11. night
12. There is a legend that they crawl into the ears of sleeping persons.
13. animals (blood), or a host
14. 2 weeks

Not insects (answers will vary):
    Spider (too many legs)
    Dog Tick (too many legs)
    Millipede (too many legs, non-segmented body)
    Wood Louse (too many legs)

pages 22–23

Answers may vary.
1. caterpillar
2. leaf/plant
3. caterpillar
4. beetle
5. beetle
6. shrew
7. shrew, snake, owl
8. makes its own food
9. beetle, shrew, snake, owl
10. caterpillar, beetle, shrew, and

snake
11–30: Answers will vary somewhat. Accept any one of the answers or other answer student can verify.
11. insects, spiders, worms, minnows
12. people, birds, skunks, snakes, raccoons
13. zebra, antelope, warthog, fish, turtles, birds
14. birds of prey, snake, fox, coyote, weasel, people
15. fish
16. cricket, ladybug, spider
17. fish, frogs
18. people, cougar
19. insects, small animals (bugs)
20. bird, frog
21. smaller fish
22. mussels, clams, oysters
23. people, seal, whale, fish
24. birds, frogs, rodents
25. lion, hyena, leopard, cheetah
26. deer, elk, skunk, porcupine
27. bobcat, owl, cougar
28. aphids
29. weasel, cougar
30. crabs, clams, snails

pages 24–25

Answers may vary somewhat. Check to see that student has the general idea.
1. carry loads of oxygen-rich blood away from the heart
2. keep pressure equal on both sides of the eardrum
3. cover trachea during swallowing to keep things out of breathing tubes
4. control and manage thinking and awareness
5. take air that has been breathed in and pass it down to the bronchi
6. clean unwanted substances out of blood
7. assist breathing by expanding and contracting to help lungs get and remove air
8. chomp and grind food
9. join muscles to bones
10. carry urine from kidneys to the bladder
11. cover and protect the body
12. keep bones from grinding against each other
13. control the body's metabolism
14. carry signals through the body
15. let light into the eye
16. break up and soften food for digestion
17. make bile to break up fats during

digestion
18. produce antibodies to fight disease, attack harmful bacteria
19. make red blood cells

pages 26–27

8:30   tetanus
9:00   chicken pox
9:30   mumps
10:00 warts
10:30 athlete's foot
11:00 dandruff
11:30 food poisoning
1:00   laryngitis
1:30   arthritis
2:00   pneumonia
2:30   snoring
3:00   bruise
3:30   flu
4:00   hiccups
4:30   sprain

pages 28–29

1. moraine—ridges of debris deposited by melting glaciers
2. abrasion—scouring action of wind-carried particles resulting in erosion of rock
3. fault—a fracture in a rock along which movement has taken place
4. landslide—the rapid downhill movement of large amounts of Earth materials
5. delta—a fan-shaped deposit that develops when moving water loses its velocity
6. erosion—the processes by which Earth materials are worn away or carried away and deposited
7. flood plain—an area of fertile soil composed of find sediment deposited during floods
8. fossils—the remains or traces of once-living organism preserved in rocks
9. sinkhole—a funnel-shaped depression in Earth's surface that results from limestone being dissolved
10. loess—a windblown deposit of fine dust particles from deserts, dry riverbeds, and old glacial lakebeds
11. canyon—channel or gorge scoured out by the action of a moving river
12. stalagmite—upward growth of calcium carbonate from the floor of a cave
13. mudslide—rapid, downhill movement of Earth materials after a heavy rain

pages 30–31
1. imposter
2. imposter
3. for real
4. imposter
5. for real
6. imposter
7. for real
8. garnet
9. for real
10. for real
11. imposter
12. imposter
13. for real
14. for real
15. for real
16. imposter
17. imposter
18. graphite

pages 32–33
1. sandstorm—heavy winds driving clouds of sand
2. blizzard—high winds and heavy, blowing snow
3. sleet—raindrops falling through freezing air causing the drops to freeze
4. hurricane warning—hurricane conditions exist and a hurricane could develop
5. hurricane—extremely high winds developing over the warm tropical ocean
6. monsoon—heavy rain from monsoon winds that develop in the Indian ocean and southern Asia
7. gale-force winds—winds between 32 to 63 miles per hour
8. tornado watch—the kind of weather exists that could develop into tornado conditions
9. cyclone—air flowing out counter-clockwise from an area of low pressure causing a major storm
10. tornado—destructive, whirling funnels of air developing between the bottom of a storm cloud and the ground
11. waterspout—a tornado over the water
12. hailstorm—drops of water frozen in layers around a nucleus of ice
13. fog—water droplets forming and hovering over the ground
14. thunderstorm—a storm accompanied by thunder and lightning
15. tsunami—a huge wave formed by earthquake action

pages 34–35
1. Mariana Trench
2. Gulf Stream
3. Off coast of Florida, Florida-Bermuda-Puerto Rico
4. to distract predators
5. pull of gravity of the Earth-Moon-Sun
6. all shoreline formations
7. lights
8. secretes slime to cover the spines
9. plankton
10. barracuda
11. Kraken
12. spits a stream of water that turns the urchin over, eats the softer underside

pages 36–37
A. meteor shower—a group of space materials burning up in Earth's atmosphere; generally particles from the break-up of a comet
B. rotation—a planet turning on an axis
C. revolution—a planet moving around the sun on an orbit path
D. full moon—moon phase where moon is fully visible from Earth
E. comet—mass of space dust and rocks, ice, and frozen gases with a long tail
F. shooting star—a briefly-visible meteor
G. meteorite—meteor that has hit Earth's surface
H. solar flares—sudden increases in the brightness of sun's chromosphere
I. new moon—moon phase where moon is not visible at all from Earth surface
J. lunar eclipse—moon moves into the shadow of Earth and is not visible
K. orbiting satellite—natural or man-made object circling around a planet or other body
L. solar eclipse—moon casts a shadow on the Earth blocking out all or part of sun's rays in an area
M. super nova—a star whose sudden increase in brightness gives a greater burst of energy than a nova
N. black hole—a condensed star with a very strong gravity field
O. white dwarf—a white star of low burning intensity

pages 38–39
1. Saturn
2. Jupiter
3. Mars
4. Mars
5. Saturn
6. Neptune
7. Mercury
8. Venus
9. Uranus
10. Pluto
11. Earth
12. Venus
13. Pluto
14. Jupiter
15. Jupiter
16. Venus
17. Neptune
18. Mars
19. Earth
20. Mercury or Venus
21. Mercury
22. Earth or Mars

pages 40–41
Answers will vary somewhat. Check to see that student has expressed the general idea of the science principles at work.
1. The cold of the freezer makes the movement of the molecules slow down, so the molecules get closer together until the water becomes a solid.
2. The sound is gas (carbon dioxide, dissolved in the soda) escaping.
3. The Doppler effect: pitch of sound is higher when an object is moving toward you because of greater sound wave compression.
4. Grocery items have a bar code (with price) that is read electronically.
5. The screwdriver acts as a lever, putting greater force on the lid of the can.
6. because of the spinning of the Earth
7. For every force, there is an opposite and equal force.
8. Water is an excellent conductor of electricity. The electricity will go through the water into your body.
9. The buoyant force of the water is greater than the force of gravity pulling down on the shark.
OR,
the density of the shark plus the air inside the shark is less than the density of the water.

10. Heat travels from the fire through the air by convection and radiation.
11. The water in the clothing evaporates.
12. The motor removes air from inside the vacuum cleaner, creating a vacuum. Then the outside air has greater air pressure than inside. The outside air rushes in, pulling in dirt with it.
13. The rays of light are bent as they pass through the water. This makes the object seem to be in a different place.

pages 42–43

1. telescope
2. electrocardiogram
3. microscope
4. gravity
5. atom
6. electricity
7. inner tubes
8. telephone
9. telegraph
10. lenses or eyeglasses
11. neon lights
12. jet engine
13. fax machine
14. electric light bulb
15. syringe
16. x-rays
17. printing press

pages 44–45

1. incorrect—potential energy
2. correct
3. correct
4. incorrect—inertia
5. incorrect—different velocities
6. correct
7. 770 miles
8. Colorado Cruiser
9. 120 mph
10. 8 hours, 12 minutes
    (or eight and two-tenths hours, or eight and one-fifth hours)
11. 98 MPH
12. 5 hours

pages 46–47

1. It absorbs all the colors in light.
2. Water drops in the clouds scatter all the waves of light to show all the colors.
3. Earth blocks the sun's rays from hitting the moon.
4. Lightning is caused by an electrical connection between a cloud and the ground.
5. Only the red waves in light are reflected.
6. Laser light is more powerful; all waves emerge from source moving together "in step."
7. The refraction (bending of light) makes the sky appear to be on the ground, looking like water.
8. The object reflects all the waves in light.
9. Dust particles in the air scatter the waves in light. The shortest waves (blue) are scattered the most and therefore are most visible.
10. Multiple reflections of an image or object appear.
11. The sun's rays are reflected off dust and water particles in the air; most of the light waves reflected are the longer ones which are red and orange.
12. Light is bent (refracted) as it passes through the water, making the part of the straw in the water appear to be in a different place than it is.
13. No light passes through it.
14. The moon blocks the sun's rays from reaching Earth.
15. Electrically charged particles from the sun strike molecules in the atmosphere and release energy.
16. Light passes through it and you can see through it.
17. Light rays are refracted (bent) by ice crystals in the upper atmosphere.
18. Light passes through it but you cannot see through it.

pages 48–49

True statements are:
1, 2, 3, 4, 6, 8, 9, 11, 12, 15, 16, 17, 19, 20, 22, 23, 24, 25, 26, 29, 31, 33

page 50

1. Static electricity has been built up.
2. Rubber tires are poor conductors of electricity.
3. This is a reflex action—the body's electrical/nervous system automatically sends signals to the foot to kick.
4. The circuit has been interrupted or broken somehow.
5. Static electricity has been built up.
6. Lightning strikes the tallest object in an area, and a tree is taller than the ground.
7. The covering insulates the water heater to reduce heat loss.